$23.00

DATE DUE


```
J
609          LUCAS, DEBRA
LUC

      ONE THING LEADS TO ANOTHER
```

Science Links

One Thing Leads to Another

Debra Lucas

Laguna Vista Public Library
1300 Palm Blvd.
Laguna Vista, TX 78578

CHELSEA CLUBHOUSE
An Imprint of Chelsea House Publishers
A Haights Cross Communications Company
Philadelphia

This edition first published in 2003 by Chelsea Clubhouse, a division of
Chelsea House Publishers and a subsidiary of Haights Cross Communications.

A Haights Cross Communications Company

This edition was adapted from Newbridge Discovery Links® by arrangement with Newbridge Educational Publishing. All rights reserved. No part of this publication may be reproduced or transmitted in any form or by any means without the written permission of the publisher. Printed and bound in the United States of America.

Chelsea Clubhouse
1974 Sproul Road, Suite 400
Broomall, PA 19008-0914

The Chelsea House world wide web address is www.chelseahouse.com

Library of Congress Cataloging-in-Publication Data
Lucas, Debra.
  One thing leads to another / by Debra Lucas.
    p. cm. — (Science links)
Includes index.
Summary: Discusses how ideas can lead to inventions including the cases of roller skates, motion pictures, windshield wipers, lightbulb filaments, and toll house cookies.
  ISBN 0-7910-7425-0
1. Inventions—Juvenile literature. [1. Inventions.] I. Title. II. Series.
T48 .L83 2003
609—dc21
                                                                        2002015892

Copyright © Newbridge Educational Publishing LLC

Newbridge Discovery Links Guided Reading Program Author: Dr. Brenda Parkes
Content Reviewer: Jenifer Helms, The Tech Museum of Innovation, San Jose, CA
Written by Debra Lucas

Cover Photograph: An early lightbulb
Table of Contents Photograph: Windshield wipers at work

Photo Credits:
Cover: Chris Rogers/The Stock Market/CORBIS; contents page: Annie Griffiths: Belt/CORBIS; pages 4–5: David Young-Wolff/PhotoEdit; page 6: (left) Bettmann/CORBIS, (right) CORBIS; page 7: Philip Goule/CORBIS; page 8: Bob Daemmrich/Stock Boston; page 9: North Wind Pictures; page 10: Syracuse Newspapers/The Image Works; pages 10–11: CORBIS; page 11: Robert Landau/CORBIS; pages 14–15: Bettmann/CORBIS; page 15: Kit Kittle/CORBIS; page 16: (top) Courtesy of the U.S. Patent Office, (bottom) Digital Image © Photodisc; page 17: Annie Griffiths Belt/CORBIS; page 18: Bettmann/CORBIS; pages 18–19: Mike Zens/CORBIS; page 20: (top) Stock Montage, Inc., (bottom) Stock Montage, Inc.; page 21: Wolfgang Kaehler/CORBIS; page 22: Courtesy of Nestle USA; page 23: Gale Zucker/Stock Boston; page 24: Bauman Photography, Inc.; page 25: (bottom) Stephen Ogilvy; page 26: Connie Hansen/The Stock Market/CORBIS; page 27: Courtesy of Nestle USA; page 28: Courtesy of Kathy Garcia; page 30: Steve Chenn/CORBIS

Illustrations by Gershom Griffith, page 4; Sidney Jablonski, page 29

While every care has been taken to trace and acknowledge photo copyrights for this edition, the publisher apologizes for any accidental infringement where copyright has proved untraceable.

Table of Contents

One Thing <u>Can</u> Lead to Another 4

An Amazing Development 8

A Clean Sweep 14

Bright Ideas 18

An Unexpected Break 22

Meet a Young Inventor 28

    Glossary 31

    Websites 31

    Index 32

One Thing Can Lead to Another

This invention wasn't an accident, but it sure created one!

4

In 1760, a violinist named Joseph Merlin invented roller skates because he wanted to make a grand entrance to a party. Modeling them after ice skates, he attached two wheels along the center of each shoe and strapped them to his feet. The crude design would not allow for turning or stopping, so when he rolled in, he crashed and broke a mirror and his violin and hurt himself. He certainly did make a grand entrance!

How are these modern skates different from the ones Joseph Merlin made?

In the early 1900s, skates still had four separate wheels and clamped onto shoes. Skating rinks and dancing on skates became popular.

Over the years, other inventors created their own roller skates. Most early skates had wheels made of wood or metal. These skates had wheels in one straight row attached to the bottom of a shoe. But unlike the in-line skates that we use today, such skates could not follow a curved path. Then, in 1867, James Plimpton came up with a more practical kind of roller skate. His skates had four wooden wheels: two in the front of the skate, and two in the back. Someone wearing Plimpton's skates could make turns and even skate backward.

Over time, new materials and designs helped make roller skating the popular activity it is today. Modern in-line skates got their start in 1979 when two brothers found an antique pair of in-line skates and were inspired to use modern technology to make the old design practical. The new in-line skates are built to be lightweight, turn and stop easily, and give the skater more control. Skateboards, another modern invention, also make use of Joseph Merlin's basic ideas.

If you stop and think about it, everything we use had to be invented by somebody. But some inventors didn't realize at the time that their creation or ideas would have such an **impact** on people's lives. Whether it's skates, or cookies, or lightbulbs, many items are here today because of a series of revisions and even accidents. Get ready to discover the true and surprising stories behind some of the things you use every day.

These skates are designed for speed skating. Why do you think that the inventor made the wheels so large?

7

An Amazing Development

Movies and television programs entertain and inform us with amazing **special effects**, computer images, and **innovative** camera techniques. It is hard to imagine the world before motion pictures and television.

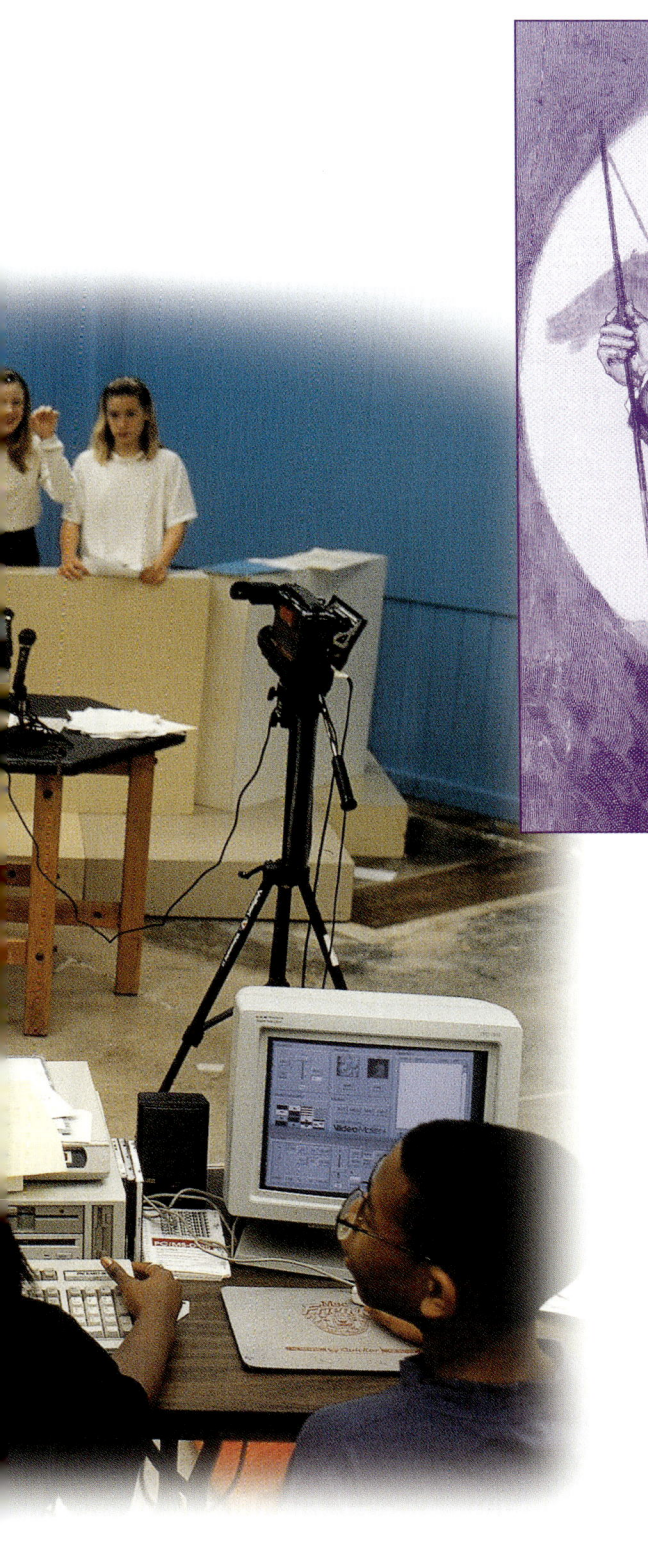

Eadweard Muybridge demonstrates his method of producing moving images.

But it was more than 100 years ago when a man named Eadweard Muybridge became the special effects wizard of his day and changed the world of pictures forever. It all started with an argument about a horse.

A video production class learns current television techniques.

9

In the 1850s Eadweard Muybridge left his home in England and came to San Francisco, California. He quickly became known and respected as an excellent photographer, specializing in **panoramas** of his new homeland.

One day, Leland Stanford, the former governor of California, disagreed with his friends about how a trotting horse moves. Stanford wanted to know whether, at any point, a trotting horse would have all four feet in the air at once. He asked Muybridge to use photographic skills to record his horse's movements.

So in 1872, Muybridge photographed Stanford's horse, Occident. Muybridge made his own camera shutters and set up 27 cameras that allowed him to take many, many photos, one right after another. In the end, viewing these photos did not quite settle the argument, but something totally new developed.

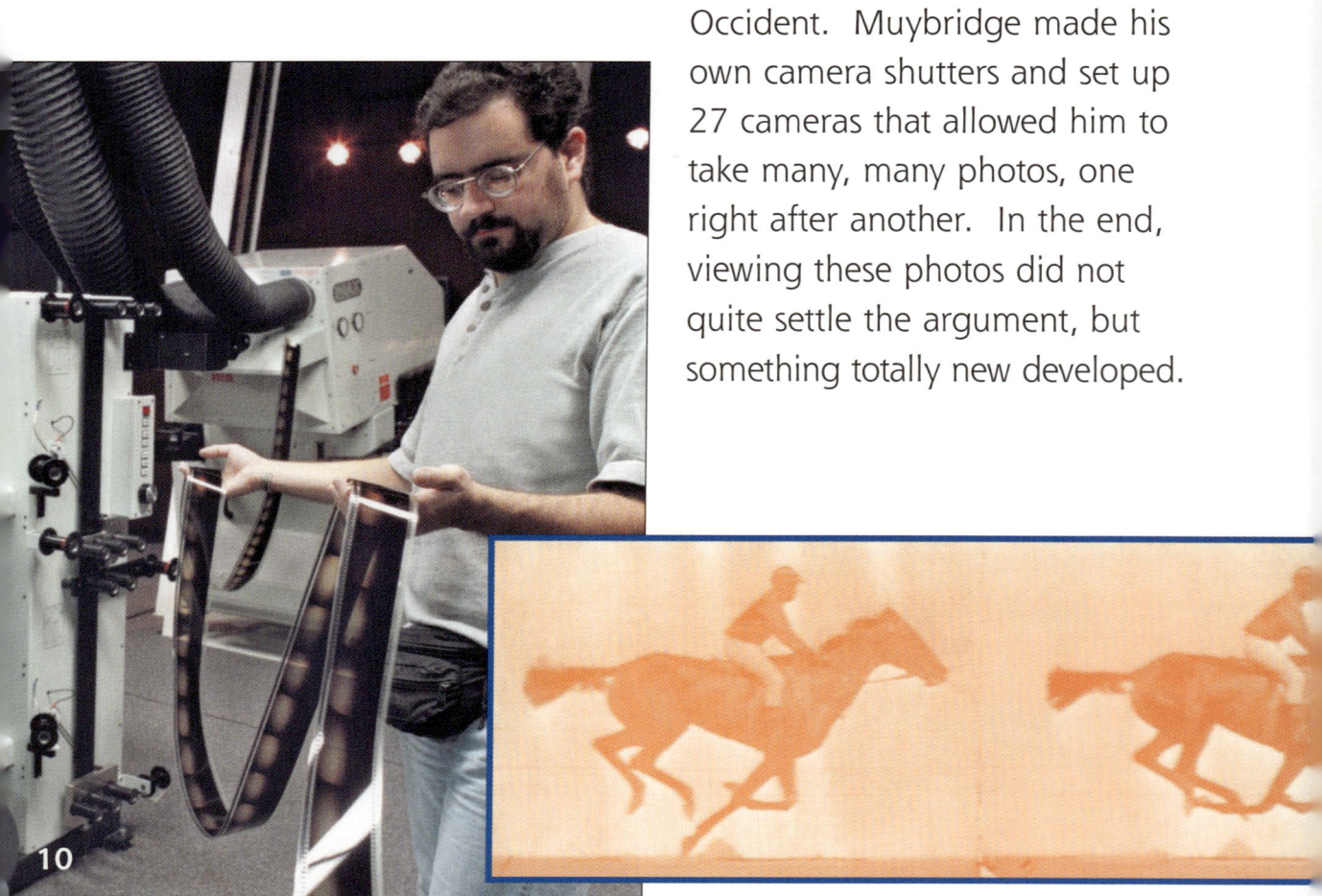

Modern motion picture cameras create a series of still images (as shown on page 10), similar to those Muybridge created so long ago.

Muybridge realized that if he could make his series of photos move, people could really see Occident in action. He went on to become the first person to use still photography to give the illusion of motion.

Later he proved that sometimes all four feet of a galloping horse are in the air at once.

Here's Muybridge's proof that a galloping horse can have all of its feet off the ground at the same time.

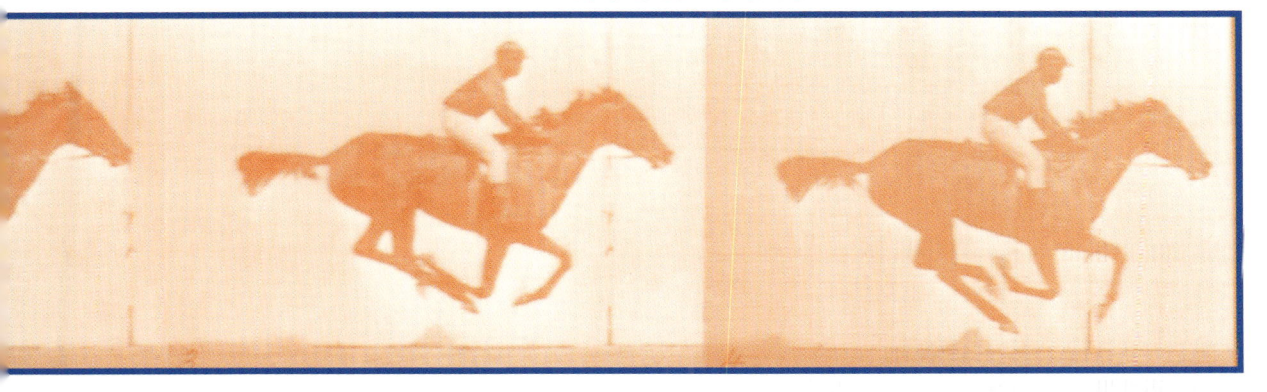

Muybridge invented a machine that projected motion. He called it a zoopraxiscope. It was a glass disk that had every image lined up around the circle. Individually, each picture was just slightly different from the previous one. But as each passed quickly in front of people's eyes, they saw a sequence that appeared to move. Muybridge used his new photographic technique to make sequenced photographs of

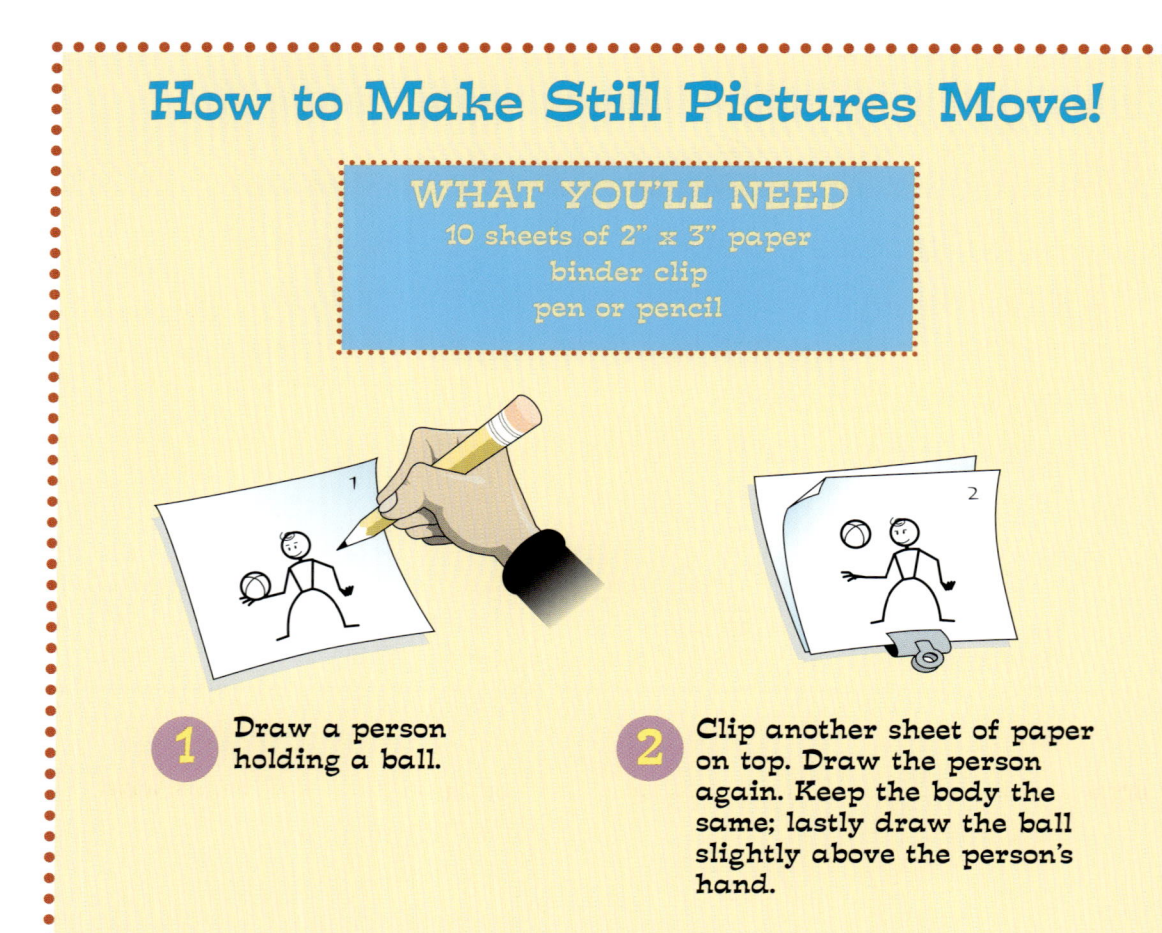

How to Make Still Pictures Move!

WHAT YOU'LL NEED
10 sheets of 2" x 3" paper
binder clip
pen or pencil

1 Draw a person holding a ball.

2 Clip another sheet of paper on top. Draw the person again. Keep the body the same; lastly draw the ball slightly above the person's hand.

other animals and of people as well. The public was astounded at what they saw.

In his day, Muybridge was considered to be a special effects genius. Today, movies remain a series of single pictures, each one slightly different from the previous, all moving very quickly in front of your eyes.

3 Add a third sheet on top. Draw another picture, with the ball again slightly higher.

4 Flip the pages quickly so you can see how the three drawings move together so far.

5 Continue to add sheets of paper, moving the ball in each drawing, first higher, then over the person's head, then moving toward the person's other hand.

6 Flip all the pages with your thumb, and watch the ball move!

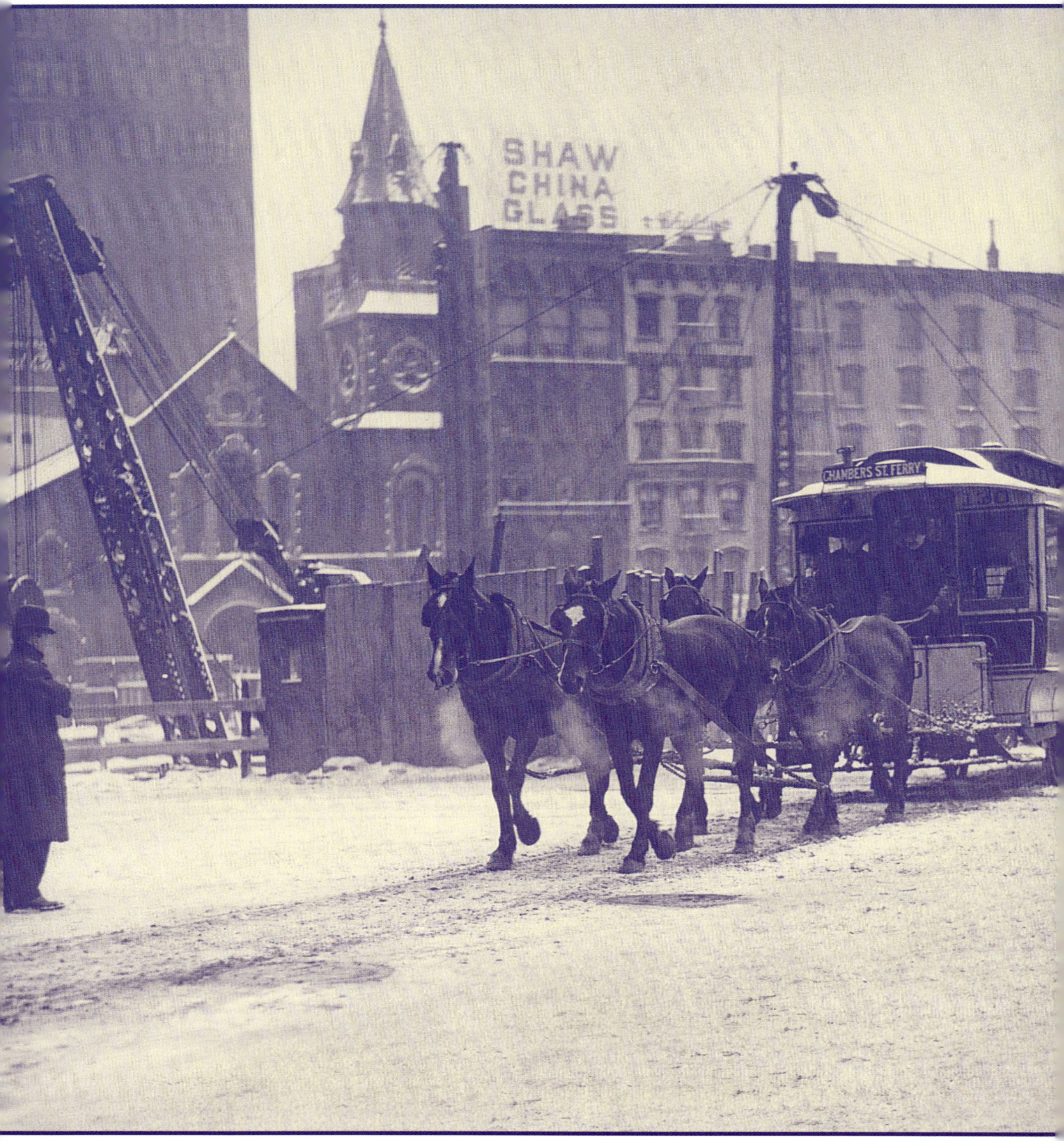
SHAW
CHINA
GLASS
CHAMBERS ST. FERRY

A Clean Sweep

It was a snowy day in 1902 when Mary Anderson took her first ride on a New York City **streetcar**. She was visiting from Alabama and wanted to see some of the sights of New York. But little did she know that her ride would eventually help millions of people.

Mary watched the streetcar **motorman** shiver as he struggled to clear the snow that piled up on the windshield every few minutes. There were no windshield wipers back then. Instead, the motorman opened up a section of the windshield in the middle and lifted it up, allowing him to reach out and scrape off the snow. Not only were his hands freezing, but he got a face full of snow each time he cleared the windshield.

Mary Anderson considered the driver's problem and knew there had to be a better, more comfortable way to keep the windshield clear.

How are today's buses different from early streetcars?

Right there on the streetcar, Mary took out her notebook and began to draw an idea for a new wiper with an attached **lever**. The lever was a bar that would be inside the car, right near the motorman. When he moved the lever, it would activate a swinging arm outside on the windshield. The arm would move back and forth, wiping the snow away, and the motorman would be able to control it from inside the streetcar. Mary designed the wiper so that it would be removed during the nicer weather, since she thought it would spoil the looks of the streetcar.

Mary's patented windshield wiper

Mary returned to Alabama with her sketches. Many people told her it couldn't be done. But Mary was convinced that

Before there were windshield wipers, people rubbed a cut onion or carrot stick across their windshield. This created a greasy film on the glass, which lessened the buildup of snow and ice.

16

streetcar and auto companies would want to buy her invention. She moved ahead and received a **patent** for her "window cleaning device." Receiving the patent meant that Mary was the only person who could make or sell her invention for a number of years. Unfortunately, she could not find a company willing to buy her idea. Yet, after Mary's patent ran out, car companies began installing their own versions of the windshield wiper. To this day, Mary Anderson's clear idea has led to something drivers can't be without. The next time you are able to see out the windshield of a car or a bus, remember that it all started because one woman felt sorry for a shivering motorman.

The first windshield wipers, made of wooden strips and rubber, moved by a hand-operated lever. Today, they work with a push of a button, and move back and forth at different speeds.

Bright Ideas

You flick a switch, or turn a knob, and a lightbulb brightens the room, allowing you to read, work on the computer, or even just find your way. But the simple lightbulb, as we know it today, has come a long way.

Although the electric lightbulb had already been invented by the late 1800s, many people didn't want to use lightbulbs in their homes. The bulbs burned out very quickly, lasting for only a few days. It was costly to repair them so often.

So, inventors all around the world began a frantic race to develop one that would last a long time.

Left, an early lightbulb
Right, Seattle glows at night.

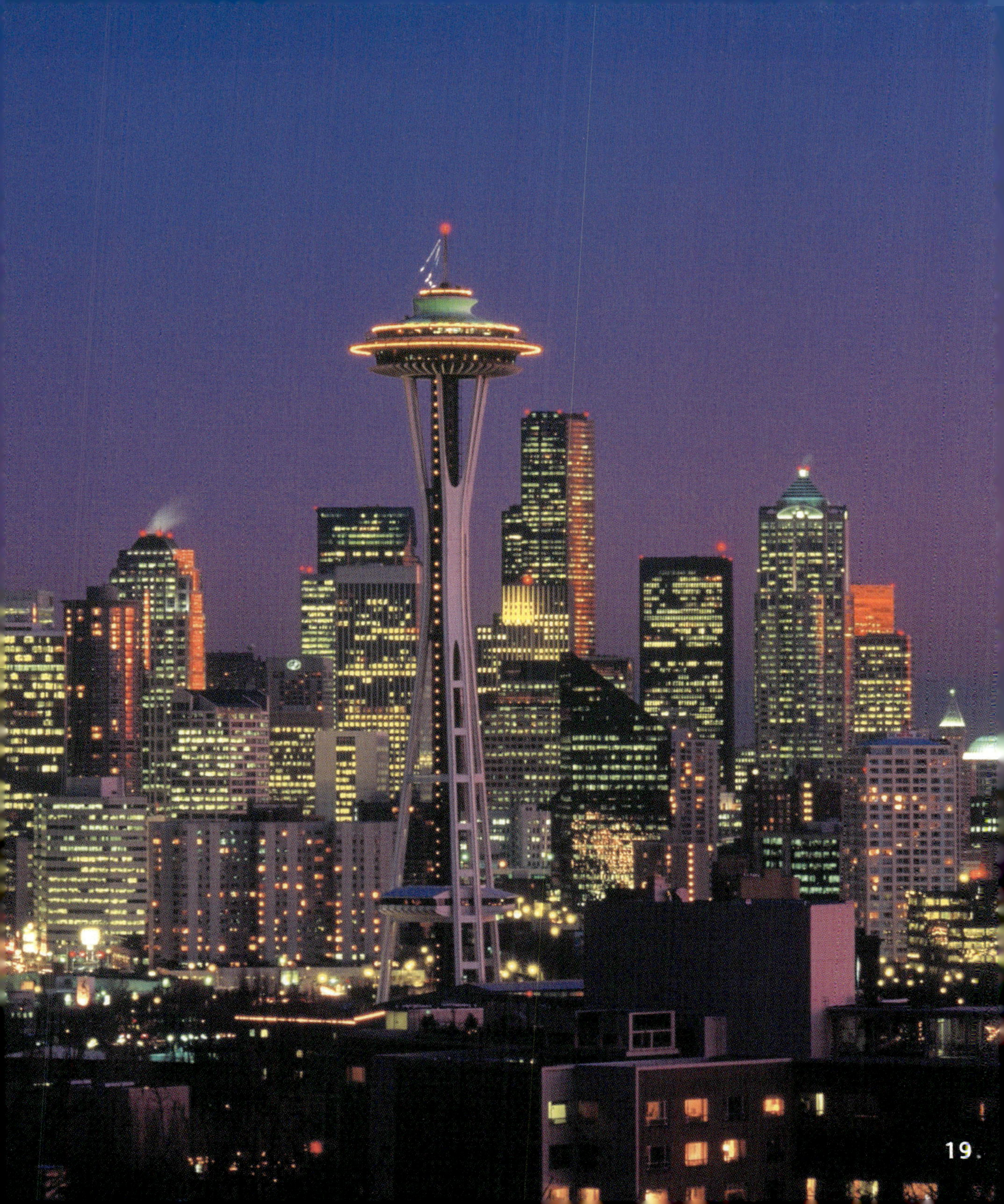

An inventor named Lewis Latimer set out to solve the problem by experimenting to find a better way to make the **filament** for the lightbulb. The filament is the part inside the bulb that heats up and gives off glowing light. But the filaments in the first lightbulbs were irregular and fragile. They did not last long. His new method of making filaments resulted in lightbulbs that were less expensive and would last longer than ever before. He used cardboard envelopes to keep the carbon fibers of the filament rigid during a process of airless heating.

Latimer applied the same problem-solving skills to dealing with bringing electric light into every home and building.

Lewis Latimer

This drawing shows the parts of the lightbulb. The filament is seen both in the bulb and as figure 2.

Complex electrical engineering was needed for powering and wiring cities. His company sent him to set up lighting systems for major cities in the United States, and even as far away as London, England. He even wrote a book that electrical engineers used as a guide. In 1884, Thomas Edison contacted him, and the two of them worked closely together on new inventions and patents. Latimer later became one of the first members of the Edison Pioneers, a special group of inventors who had been the early leaders of the electrical industry.

Later, inventors improved the durability of the lightbulb, but who knows? Without Lewis Latimer, we might all be in the dark.

How is the modern lightbulb like Latimer's?

An Unexpected Break

When one imaginative woman tried something new, an unexpected result led to a whole new breakthrough in eating: chocolate chip cookies.

In 1930, Ruth and Kenneth Wakefield bought a house that had been a busy stopover for people traveling from Boston to New Bedford, Massachusetts, since the 1700s. It was called a **toll house**. Here travelers paid a fee (or toll) for using the road, changed horses, and could get a home-cooked meal. The couple worked hard making the place into a modern-day rest stop.

Ruth wanted the food she offered her guests to remind them of the inn's long history. She searched out old recipes. One of her favorites was for making a kind of cookie called Butter Drop Dos.

The Toll House Inn

Today, almost everyone knows how to make chocolate chip cookies, but, before 1930, no one had ever heard of them.

One day, Ruth decided to add a new twist to the old recipe. She broke a bar of chocolate into small chunks and stirred the chocolate into the cookie dough. She expected that the candy bits would melt, creating chocolate swirls through the cookies. But that wasn't what happened.

From left, Ruth Wakefield, Duncan Hines (an early food critic), and Kenneth Wakefield

　　The chocolate didn't melt the way Ruth thought it would. Instead, chunks of chocolate remained in the cookie. What would her guests think?

　　It turned out that people loved the cookies! Ruth made them for her guests over and over again.

　　People began asking Ruth for the recipe for her delicious cookies. Her recipe was even published in a Boston newspaper. Soon after, other newspapers in the area published the recipe, too.

The Nestlé company noticed that its sales for semisweet chocolate bars were skyrocketing in that part of New England. Puzzled, company officials sent an investigator to find out why. They were stunned to learn that the reason for the boost in sales was that so many people in and around Boston were buying chocolate in order to follow Ruth's recipe.

Delighted with this new use for its product, Nestlé decided to put **score lines** in their chocolate bar to make the chocolate easier to break into chunks. The company also offered a special chopper to cut the chocolate.

What's Your Favorite Cookie?

Here's how Americans answered this question in a poll conducted in 2001:

- Chocolate Chip 60%
- Oatmeal 16%
- Other 13%
- Peanut Butter 11%

These chocolate chip cookies became known as toll house cookies. Nestlé bought the name from the Wakefields and printed their recipe on the back of their chocolate bars that were used for baking. As part of the deal, Nestlé also supplied Ruth with a lifetime supply of chocolate to use in her baking.

Soon, packaged chocolate chips were created and people didn't have to break up the chocolate anymore. They could just pour the chips right out of the bag.

Ruth Wakefield's invention is still popular today. You can make cookies like hers, too, using any kind of chocolate you like!

To honor Ruth Wakefield and her invention, Massachusetts named the chocolate chip cookie its official cookie in 1997.

Chocolate Chip Cookies

2 cups all-purpose flour
1 teaspoon baking soda
1 teaspoon salt
1 cup butter or margarine, softened
2 cups sugar
1/2 cup brown sugar, firmly packed
1 teaspoon vanilla extract
2 eggs
One 12-ounce chocolate bar cut into chunks or 12 ounces of chocolate chips

Preheat oven to 375°F. In a bowl, mix flour, baking soda, and salt. In a larger bowl, combine butter, sugar, brown sugar, and vanilla extract; beat until creamy. Beat in eggs. Gradually add flour mixture and mix well. Stir in chocolate. Drop by rounded measuring teaspoonfuls onto ungreased cookie sheets. Bake at 375°F for 8-10 minutes.

Meet a Young Inventor

You have read about how mistakes or discoveries can eventually evolve into something new. Now meet a teenager who set out to solve an everyday problem in order to create an award-winning invention.

Jennifer Garcia, of Newark, New York, was in the seventh grade when she invented her Vacuum Dirt Mat. Her mother kept asking her to vacuum the dirt that was constantly being tracked into the house from their unpaved driveway. It was tiresome for Jennifer to keep cleaning up the gravel and dirt every time she came inside. A regular floor mat didn't seem to clean her shoes very well, and she wondered how she could improve upon it. This led Jennifer to the idea of vacuuming the bottom of her shoes, instead of vacuuming the entire floor! This was the first, and very important, step to making her invention work. She had identified her problem and come up with a partial solution. Now she could take steps to make the invention reality.

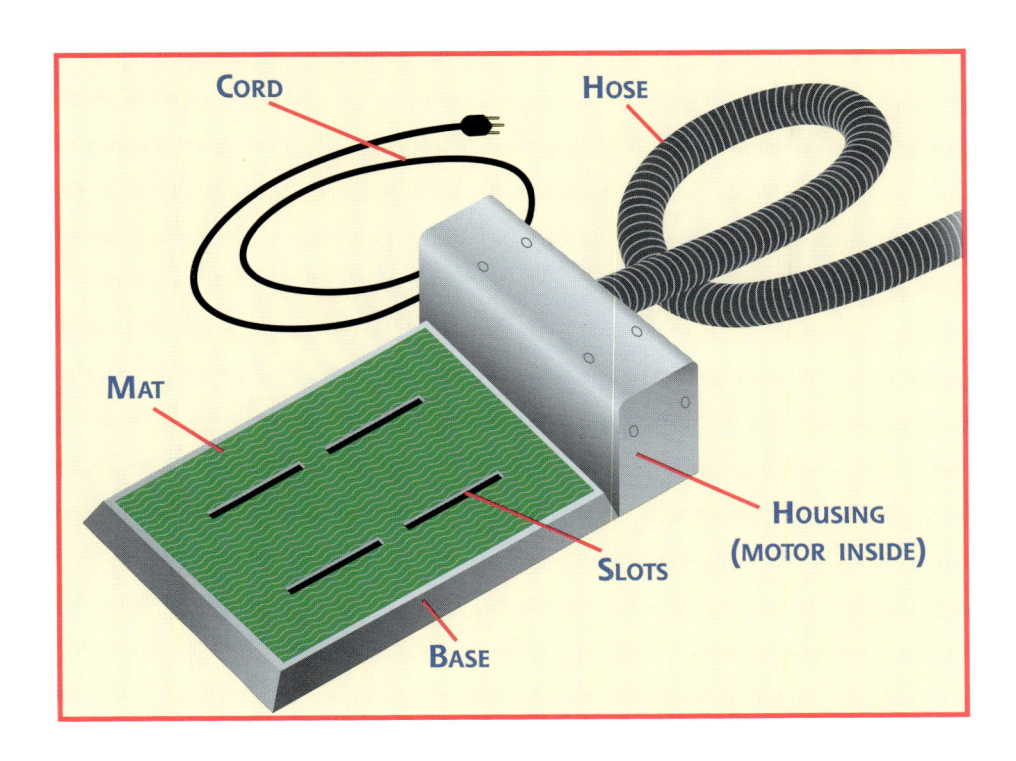

So Jennifer set to work on developing a floor mat that had a vacuum attached to it. First she drew a plan. It was a picture of how the mat would look. She added all the parts that would have to be constructed. Her drawing included a vacuum hose, the electrical cord, plug, the mat, and a wooden platform. Next, she began building her invention.

Jennifer built her vacuum a section at a time. She would work on attaching the vacuum one day, and making the slots in the mat the next day. "By taking each section at a time," she said, "you know exactly where your problems are." Finally, the Vacuum Dirt Mat was finished. It won first place in New York State's Invent America! competition.

Today, Jennifer is in college, studying medical technology, and hopes to go into the field of medical research. Certainly, finding new and creative solutions to problems is something she can do well. Maybe someday Jennifer will come up with a valuable new medical breakthrough, which in turn may lead to even more discoveries. After all, one thing does lead to another.

This boy is inventing a new structure. You probably analyze problems and invent solutions often. You can be an inventor, too!

Glossary

filament: thin wire in a lightbulb that electricity causes to glow

impact: a strong force or effect

innovative: creative; making something never seen before

lever: a rod people use to move something else

motorman: an operator of a subway or a streetcar

panoramas: views of an area that take in all directions

patent: government papers giving inventors exclusive right to their devices for a period of time

score lines: straight cuts or marks on a flat surface

special effects: sounds or images used to create illusions in movies or on television

streetcar: a street vehicle that transports people and which is still found in some cities

toll house: a house on the side of the road where tolls were collected

Websites

Find out more about inventors and inventions at
www.uspto.gov/go/kids/kidtwink.html
www.pafinc.com/students/siba_winners.htm
www.inventorsmuseum.com/museum_map.htm
kids.patentcafe.com/

Index

Anderson, Mary 15–17

automobiles/cars 17

buses 15, 17

chocolate chip cookies 22–27
   recipe 27

cookies, favorite kinds,
   chart 26

Edison, Thomas 21

electric lights 18, 20, 21

filaments 20

flip book, instructions 12–13

Garcia, Jennifer 28–30

horses 10–11

in-line skates 6–7

Latimer, Lewis 20, 21

lightbulbs 18, 20, 21

medical technology 30

Merlin, Joseph 5, 7

motion pictures 8–13

Muybridge, Eadweard 9–13

Nestlé Chocolate Company 25, 26

Occident 10, 11

patents 16, 17, 21

photography 10–13

Plimpton, James 6

roller skates 5, 6, 7

skateboard 7

Stanford, Leland 10

streetcars 15–17

toll house cookies 22–27

Toll House Inn 22

Vacuum Dirt Mat 28–30

Wakefield, Ruth 22–27

windshield wipers 15–17